The Law of Science versus the Law of Life
Nuclear Accidents and the Limits of Human Control

A Lecture by Jinzaburo Takagi

Published by Octave Publications
Ichijoji Matubara-cho
Sakyo-ku, Kyoto, Japan 〒606-8156
Tel. +81-75-708-7168

ISBN4-89231-136-9

Printed in Japan
November 2015
First Edition

Table of Contents

Preface

On March 26, 2011, the Citizens' Nuclear Information Center (CNIC)[i] received a letter from Akira Yukimura, director of the Kanazawa District Doctrinal Study Group of the Japanese Shin Buddhism Association, in regard to a lecture given by my husband, Jinzaburo Takagi, on February 22, 1991, at their district meeting. The lecture was titled, "The Law of Science versus the Law of Life," and Mr. Yukimura indicated that, since it dealt with an issue of universal importance, he wished to compile it into a booklet for publication.

At the time, however, the CNIC team was busy responding to ongoing problems that arose from the Fukushima Dai-ichi nuclear disaster triggered by the Great East Japan Earthquake[ii] that shook Japan on March 11, 2011 so Mr. Yukimura's request was forwarded to me.

My husband, Jinzaburo, died of cancer in October 2000 at the age of 62. It would have been ideal if he could have read the proofs of this booklet himself but that was obviously not possible. Instead, I agreed to proofread the transcript that was quickly prepared since I often helped him with his lectures and performed audio transcription for him during his life. I sincerely hope that this booklet will be read by many.

As he mentions in the text, only a few days before the lecture, on February 9, 1991, a steam generator pipe ruptured at Unit 2 of the Mihama Nuclear Power Plant[iii] operated by the Kansai Electric Power Company. This rupture resulted in the leakage of tons of primary coolant which activated the emergency core cooling system（ECCS）. This incident very nearly ended in a meltdown.

Now, one month after the March11 earthquake, we are still getting aftershocks of intensity levels 5 and 6, and the situation at Fukushima Dai-ichi Nuclear Power Station is so serious that no one can predict where it might lead. This morning, it was reported that the Nuclear and Industrial Safety Agency, overseen by the Ministry of Economy, Trade and Industry, had raised the accident assessment to Level 7, the worst on the International Nuclear Event Scale; this puts the situation on the same footing as the Chernobyl disaster. Let me end this brief preface by quoting a part of the last message Jinzaburo left for his friends before he passed away: *Since I will be gone before any of you, what worries me the most in this era of nuclear power is the potential for major accidents caused by terminally unresolved risks, and a concern that radioactive waste will end up leaking into the environment. I only hope that other people will have the sense to learn from history and, having faced up to reality, will take positive actions to bring this nuclear age to a close as soon as possible. From wherever I am, I'll certainly be watching this action with interest.*

Kuniko TAKAGI

April 12, 2012

The Law of Science versus the Law of Life

Jinzaburo TAKAGI

An open lecture organized by the Kanazawa District Doctrinal Study Group of the Japanese Shin Buddhism Association and delivered on February 22, 1991.

PART I

Good evening.

Usually when I am invited to give a lecture, I talk about my areas of specialization, namely, nuclear power stations and the nuclear fuel cycle. In the past I have often been asked by Christian organizations to talk about the limits of science and technology in reference to life and humanity. But, this is the first time that I am addressing similar issues to an audience knowledgeable about Buddhist teachings and classics. Today's lecture deals with similar underlying issues, but since I was asked to talk about the relationship between the law of science and the law of life, I have taken a slightly different approach. Please bear with me while I lay out some groundwork.

The Law of Science and the Law of Life - Pushing Ourselves to the Brink of Crisis

I would like to understand the law of science primarily through the logical principles that regulate the production of scientific knowledge. In contrast, the law of life can be approached through the natural and moral laws governing our life or living. Here, I use the two phrases, science and life, in a kind of contrastive way, but, come to think that science is also manmade. Although the phrase "science and technology" is used as an idiom today, "science" and "technology" used to be independent concepts. Science was the pursuit of pure knowledge while technology was the utilization of materials and manufacturing techniques for a concrete purpose. Science used to mean an entirely different undertaking than technology. Nowadays, however, scientific discovery is often driven by certain targeted technological achievements, that is, scientists conduct their research with technical applications in mind right from the beginning. Science and technology have thus become inseparable. Historically, we can probably trace the roots of this entanglement to Europe's modernity. In any case, science and technology were originally pursued on a purely pragmatic basis, whereas, as you already know, today they are no longer subordinate to this original intent. Science used to be just one of many activities undertaken by human beings. Even in terms of learning and the pursuit of knowledge, the study of science accounted for only a fraction of our efforts and used up only a modicum of our resources. As a result of the application of scientific findings, however, what we now call "science and technology" has developed in an aberrant fashion, even to the point of making us forget that we ourselves are living creatures and therefore are still subject to the laws of nature. Moreover, as a tool of both political power and capital, science and technology has been manipulated for the purposes of waging war and generating vast sums of money. Looking at what is going on in the Gulf War[iv], I cannot but think that human beings have created something terrifying as a result of the excessive development of science and technology; we have thrown the entire natural world into crisis. As living beings that belong to and form part of the natural world, human kindare also in crisis.

I do not think it is too much of a stretch to say that we are now in a situation where we are actually endangering ourselves with our own creation, we are in the midst of a collective suicide.

I want to address this concern by looking mainly at issues surrounding nuclear power plants, with which I have personally been involved. But I will try to avoid sensationalizing the issue by describing these plants as big and frightening thingsto terrify us. If time permits, I will offer my thoughts on what we should do in light of the situation before us. First though I want to make it clear that I developed the ideas that led me to be so deeply involved in the anti-nuclear power movement through personal experiences, and not because I set out with a firm ideology or belief, not because of what I had been taught. So let me start with my personal history to help you understand where I am coming from. I should add that I might repeat some things that I've already written about in my books[v].

About 30 years ago, as a university student, I studied nuclear chemistry, which deals with radioactivity. Nuclear chemistry is the same area of study as Marie Curie worked in. It was also the field of the German chemist Otto Hahn, who discovered nuclear fission. In a broad sense, it is related to nuclear power. But my threefold focus was on whether radioactive substances produced in the course of burning fuel would leak or not, how to measure the amount of such leakage, and how these substances behave. This is somewhat different from the main areas of nuclear engineering necessary for designing nuclear reactors. After graduation, I joined a company and started researching radioactivity there.

That company does not exist anymore but since there is no reason to hide my personal history, I ought to tell you that it was Nippon Atomic Industry Group Company Limited. (NAIG); most of you have probably never heard of it. It was actually bought out last year. By the way, I think this buyout is one clue that Japan's nuclear power industry is already starting to shrink, though it may appear on the surface that it is flourishing. There's talk of a new nuclear power plant soon to be built in Suzu on the Noto Peninsula; this makes it appear that the nuclear power industry in this country is expanding. But the truth is, internally things are beginning to collapse.

You probably do not realize it but Nippon Atomic Industry Group was basically the de facto nuclear power division of Toshiba. When Toshiba entered the market, it created this new company to avoid any risk to the Toshiba Corporation itself. I joined the company very early on, about two and half years after its establishment, in the second wave of recruitment. There was not a single nuclear power plant in Japan at that time, although nuclear power generation was about to be introduced in the form of the nuclear-powered ship, Mutsu. Plans for the ship were already on the table, and my company was involved in them. But there were as yet no plans for a nuclear power station. Later, NAIG became involved in nuclear power plants in Fukushima, but I will leave that topic for now.

Anyway, that is how it was back then. The sanitized version is that we were doing basic research in preparation for the introduction of nuclear power plants to Japan. To be blunt, however, what NAIG was created to do was to conduct all kinds of research to prove that nuclear power was safe for Japan. In other words, the company was verifying various data in order to be prepared for an accident. To put it even more bluntly, it was manufacturing evidence to back up the claims it wanted to make. That was the kind of company I joined.

Back then, I never thought that nuclear power itself could be harmful to people. In that sense, I was slow to develop my ideas on this issue. I was naïve and singularly devoted to my research. Nothing more, nothing less. I went to work and day after day I wrestled with radioactive substances. About half of my time was spent on-site at Toshiba. For roughly 50% of that time I worked on the NAIG reactor, which stood in a corner of the Toshiba premises. If I remember correctly, it was the fourth or fifth reactor in Japan. Two or three had been built on the site of the then Japan Atomic Energy Research Institute (JAERI).

Toshiba and Mitsubishi had one each and Hitachi got one shortly after.

Once a nuclear power plant is built, burning the fuel generates nuclear material. This material, sometimes called "death ashes," is so toxic that one small pellet of it could kill tens of thousands of people, and more and more of this material is being produced every day.

Nuclear material is not supposed to leave the reactor. In the case of Mihama, however, nuclear material did escape. (On February 9, 1991, a steam generator tube ruptured at Unit 2 of the Mihama Nuclear Power Plant, leaking 20 tons of coolant and activating the Emergency Core Cooling System.) If I have time left at the end, I would like to talk about this accident because I'm sure it is of interest to you. In any case, the requirement is that there should be absolutely no leakage when it comes to the safety of nuclear power generation. That is why there are numerous built-in protections to ensure safety. The electric power companies call this kind of protective design a "quintuple-barrier structure[vi]."

Anyway I was working hard on my research since little was known back then about basic issues such as what happens to nuclear material when there is a rise in the temperature of the pellets inside the fuel rods and what chemical form this nuclear material takes. Because I was working right next to the reactor, I was exposed to a huge amount of radiation. You cannot even imagine how many radioactive substances I was handling; in nuclear fission rods containing regular-sized fuel, pellets produce excessive radioactivity, smaller pellets are used so as to reduce the level of radioactivity. First, the smaller pellets are blasted into the nuclear reactor via a pneumatic tube. They are then activated for a certain amount of time—minutes or hours—and the activated pellets are then blasted back with compressed air.

At this particular moment when the activated pellets are blown back, the Geiger counter does not produce any sound because the measurement goes well beyond its range. So this is territory where the Geiger counter does not work. Normally, the counter produces a sound when it detects radiation, but if the radiation is too strong, it beeps once or twice and then goes quiet. So, a higher range is set and the counter beeps again, and then goes quiet again. Again reset the range is reset, but the same thing happens. Finally, the counter goes completely dead. Working in that kind of environment, like the Geiger counter, the worker also becomes desensitized. That is just how it is for scientists and engineers. Their view of the world is mediated by machines; if the machines stop working, they become numb. Even if the radiation is strong, they do not mind.

That is the kind of place I worked in every day. Dealing with strong radiation makes you tired, and you know that you're not supposed to inhale or touch radioactive substances, so you wear a radiation protection suit and double-layer gloves—no way are your hands left unprotected—as well as lead glasses to protect your eyes and a heavy-duty mask. Thus equipped, you start to sweat like crazy after about two hours, even in an air-conditioned laboratory. But the sweat has nowhere to go, so you feel even more tired. On top of that, you are mentally worn out. However, being in a professional environment, you don't want to get beaten by radiation. It is strange, but you develop a weird sense of professional determination. When a youngster joins the team and you see how scared he is, you have to tease him a little. You tell him that if he cannot handle it he will never be a man.

That is just the way it is. In working in this kind of environment, safety is in mind. But because of this professionally developed indifference to the questions of safety, there are many problems. It may sound vulgar, but taxi drivers are often the wildest drivers, although they are usually expected to be the safest ones.

The reason is that they must earn money. It is the same case in our industry. To become more efficient, you cannot be scared all the time, and the result is numbness.

However, as someone working in the nuclear industry, the thing that originally caused me to become critical of nuclear power was not the risks posed by radiation. During my four and half years with NAIG I came across a number of work-related questions. Finally these questions compelled me to leave the company.

Of particular concern to me was the invisibility of radioactivity. Radiation is something that you can see only with the Geiger counter. Not only is it invisible, but even a speck could cause serious trouble; it is very difficult to solve the problem precisely because it is invisible. To give a trivial example here: suppose you put radioactive substances in, say, a medicine bottle or beaker, and cover it with a lid. It is unquestionable that the lid is not contaminated, but you go to the lab the next day, remove it with a bare hand without thinking, do your job, and then check with the counter for contamination before leaving the lab, and you hear a Geiger counter noise. That noise indicates the trouble. If you hear it, you must wash your body thoroughly, but it's not easy to wash off all the substances. Once they touch your skin, ordinary soap is not able to remove them.

Radioactive substances really tend to crawl outside the bottle. Usually, this is hard to believe, but it's true for some types of radioactive substances. This phenomenon and many things we still don't understand.

Capitalist Rationality and Human Concerns

In being a scientist, if I come across a puzzle, I want to solve it. If I have a question, I want to answer it. At that time, my thought was that if even one question remained unanswered, it was simply not possible to guarantee safety. But of course, my company would not let me set about working on the questions of radioactive safety. The usual procedure is as follows: when a company receives an order from, e.g. from the Tokyo Electric Power Company, it draws up a plan to make whatever it is in such-and-such a way by such-and-such date. It then adheres doggedly to that schedule. If you dare to engage in something that would necessitate an extension of the schedule, you are instantly treated like a nuisance.

Now, this kind of practice does not reflect the logical approach of science. Rather, it is driven by the capitalist rationality of the company. Some kind of force works to appropriate only those parts that are convenient for the company and to discard those that are not. Created data convenient for the company will be instantly accepted and published, while inconvenient data will simply be eliminated or ignored.

I could not tolerate this situation; it became unbearable for me. The way in which nuclear power was promoted disgusted me, rather than any intrinsic value it held. That was how I first felt alienated from the nuclear industry.

Next how did I resist the company's appropriation? How did I try to get my point across? How did I try to defend my position? The only way to do it was to justify my argument through experimentation in which the data I produced would justify my prediction. But to do so I would need measurement devices. The better the measurement device, the better the outcome I could achieve. But this line of thinking made me totally dependent on the equipment. To get a good machine, you must secure the budget. To get the budget, you have to avoid speaking out, no matter whether you work for private corporations or at universities. This is how many researchers lose their voices and become accomplices. Eventually, they just become enslaved to the industry machine.

There are lots of people working in the natural sciences

who are unafraid to stand up to their superiors. But they quake at the thought of their equipment or research tools being denied. The more committed they are to their research, the more this applies. So if their employer presses the point, many capitulate and try their best to fit the mould of the loyal employee.

I mulled over these issues for a couple of years and eventually came to realize n that my personality would become very twisted if I continued to work in a place where there was a constant threat that if my equipment were removed would I speak out. It ought to be taken for granted that I should act as a human being first and a scientist second. As soon as I present myself as a scientist, however, I stop speaking as a human being. All my actions are bound by the requirements of a scientist; my argument has to be backed up by measurements or other data. As a scientist, I have to follow the rules of the system I work for. To make a broad generalization, it is a system that trivializes people like myself within it.

At that time I was still in my mid-20s. I hadn't studied much philosophy and knew little about religions so I was not in the habit of critically assessing social situations for myself. But everything was just too restricting; I could hardly breathe. I felt as if the scope of my life had narrowed.

Finally I just decided to leave the company where I had worked for about four and a half years. By coincidence at that time, the Center for Nuclear Study at the University of Tokyo, which was conducting basic nuclear research, was advertising a position. I wanted to know more about the basics of radioactivity, like the principles of radioactivity and how radioactive substances act, so I applied for the job and got hired. There I started my new career as a university researcher; , I felt much relieved because the Center was focused on basic research. At NAIG (Nippon Atomic Industry Group Company Limited) , many of the questions debated today about nuclear power plants had yet to be raised publicly. I had always had the feeling, however, that someday they would be openly discussed as social issues. I was not absolutely against nuclear power, but I had a suspicion that the results of my research had thwarted my intention, so I had felt both conflicted and implicated in the misuses of scientific knowledge.

I had always felt two contradictory tendencies, and it was not easy to reconcile my inner conflicts. So, when I became a university faculty member, I was relieved from that kind of inner conflict. When I look back, those first two years at the Tokyo University Center for Nuclear Study were maybe the happiest of my life. I was engaged in basic research, involving the genesis and mutation of transuranium elements. The research I was involved in was also related to the field of radio-chemistry. These elements are mostly synthesized (nucleosynthesis) inside stars. When a star cools down and becomes a planet, the elements form its core; all the elements we find on earth today were created in this way.

To find out how such elements are produced, we examined the history of the Earth and collected basic data. We also examined meteorites and analyzed the findings, and so on. When we made these investigations, we also dealt with radiation. Just like in my previous work, I was once again involved in an academic discipline not unrelated to radioactivity. Of course, we did not measure strong radiation such as one would near a nuclear reactor. Nonetheless, we measured radiation even though it may have been very weak. For instance, we used to extract from rocks radioactive substances that were hundreds or thousands of years old and measure radiation from such substances using an extremely refined instrument. It was very unworldly work indeed, not something that was likely to have any practical impact on day-to-day life. We also collected mud from the seabed. For this we relied on special instruments. From a boat we lowered the bottom sampler thousands of meters down attached to wires. We then analyzed the radioactive substances contained in the ancient sediment that we had collected to find out more about radiation from space and how it affected the Earth's evolution. At that time, I had no idea that what I was doing would actually become important in a public sense.

By this time, I was in my late 20s. Sometimes I was out at sea and sometimes I was in the mountains. One day I realized that radiation was everywhere. What I wanted to measure was radiation produced tens of millions of years ago, but before I could get to it, there were radioactive substances that were only a couple of years old, and they were obstructing my investigations. I could barely do my job. It is said that a good place to collect marine sediment is the South Pacific Ocean, where there is little volcanic activity. In Japan, because of the considerable volcanic activity in the Sea of Japan and around the Japanese archipelago, the volcanic ash accumulation is relatively new, so it is of no use, not to mention the huge amount of river sediment. The place least affected by these factors is the South Pacific Ocean. There are places there where it is said to take 1,000 years for one millimeter of sediment to accumulate. If one millimeter equals 1,000 years, then digging one meter gives you 1 million years' worth of data. The bottom layer of a core sample from a one-meter-long rod explains what happened a million years ago. If the rod is 10 meters long, you get data from about 10 million years ago, and so on. This being the case, even in the South Pacific region the amount of radiation on the surface layer is exceptionally large. In other words, you can clearly see from the geological record when human beings started producing radioactive substances artificially.

In this, of course, I am referring to the time-period starting with Hiroshima and Nagasaki or, to be even more precise, the first series of nuclear tests conducted on Bikini Atoll. In that part of the Pacific region, countries like the US, the UK and France have long conducted nuclear tests, and traces of them are right there on the Earth's surface. All of this I certainly knew about in theory. When the Bikini tests happened, I was in my last year at junior high school, so I actually lived through it. And my professional forbears actually performed analyses of the nuclear fallout from Bikini. So I knew about it in an abstract way. But when I actually measured the radiation myself, even though that was not my main purpose in the South Pacific, I was astonished. I measured all sorts of things at various places in Japan as well. And everything was contaminated with artificial radioactive substances. Of course, the amount was not lethal. Just because you are near these contaminated substances, it doesn't mean that you will be affected immediately, although there may well be some long term impact. If somebody says, "It is no more than one thousandth of the tolerance dose of radiation," then I can only agree.

In any case, everywhere I tested was contaminated. This seemed to symbolize how sinful Humanity was and I was surprised in many ways. I had considered myself a successful young scientist. I had published quite a few papers by this time and I was beginning to make a name for myself. But I had known nothing about the amount of radioactive materials, or manmade radioactive substances contaminating the Earth. I was surprised that nobody around me in my area of specialization, including nuclear chemistry, knew how extensively our environment had been contaminated with radioactive materials. We experts in my field were not only ignorant of radioactive environmental contamination, but we did not care about it! That got me to reflecting. I was astonished at my ignorance as a scientist. I did not know what to do, but what I experienced subsequently proved a crucial turning point in my life.

This turnaround did not come because I detected radiation in the sea or in the mountains, but because I detected it in human dwellings. The radiation detected in there was mostly attributable to nuclear tests conducted by the US and then the Soviet Union. These two countries both conducted quite a number of tests. Later, France and China got in on the act. These nuclear tests released radioactive material into the atmosphere. Moreover, even before the nuclear testing, contamination had already been caused by isotopes used at factories. Factories must have been the setting, since there were hardly any nuclear power plants in operation back then.

This was the so-called "peaceful use" of nuclear power. However, I don't think there is such a thing as a use of nuclear power that is actually peaceful. People live in villages and towns in the countryside; there run rivers and streams inhabited by fish; flowers blossom in season and birds fly around day in and day out; these are ordinary environments where many creatures live. What was truly astonishing for me was that all these places and creatures were contaminated by radiation. Technically speaking, the level of radiation was not serious; it was only a hundred-millionth of the level of radiation I was exposed to while I worked in the laboratories at the nuclear power company. As I said before, it frequently went beyond the range of the Geiger counter at my company laboratories. In a natural environment, the level of radiation only produces a slight scratchy noise.

The funny thing is that I was insensitive to radiation in the laboratory and not surprised by it at all. But, when I detected radiation outdoors, it really disturbed me. I kept asking myself why I was astonished at the discovery of radioactive contamination in human dwellings and the natural environment while I had been totally indifferent to the much higher exposure to radiation at nuclear laboratories. I must say that in the laboratories I was totally a scientist, totally a professional. In a professional scientific environment, one gets accustomed; radiation measurements are no more than numerical figures. It was not "me," as a whole human being, dealing with radiation there. But when measuring radiation outdoors, it becomes an entirely different matter.

When you go outdoors on fieldwork, people take care of you, talk to you, and invite you to stay at their homes. And then you detect radiation in the places where they live, and they ask you, "Is it safe?" In that context, for the first time, I was confronted with radiation, not as a scientist but rather as a person, not as an isolated professional but rather as a human being living together with other human beings. Things looked totally different after that. "What in the world have we done!" This kind of question overwhelmed me. I had to think about where I stood and what I was to do. From the viewpoint of my disciplinary specialization, I would be expected to promote nuclear technology, produce more and more radioactive materials, and propagate the idea that nuclear power would contribute much to the welfare of human kind. The science I specialized in requires scientists, like myself, working in the field to offer positive publicity for nuclear power and to encourage the public to rely more and more on it. We professionals firmly believed that the universal cause for which we worked was the progress of humanity. I had never doubted for a second that this was all part of the march of progress. At the same time, I was aware that each of the scientists was responsible to produce reliable and accurate data so that our knowledge could be used safely.

I participated in the manufacture of radioactive substances since I believed that they were ultimately useful for people. But the substances we helped produce were disseminated and returned to our natural environment, back to the very sites where I was measuring the levels of radioactivity. Of course, I was not sure whether the radiation I could detect in our environment was a direct consequence of my own work. Regardless of whether I was directly responsible for the radiation contamination of our environment, I was no doubt implicated in the series of acts in which our natural dwellings were polluted by radioactive substances. However insidious it may be, however, no one can be directly blamed. Direct responsibility for the leakage of radioactive substances may lie with the companies that used them. But then there are also the people who created them, myself included. The progress of nuclear sciences had already caused the radioactive contamination of the environment. These seeds had already been sown in the field of science that I was pursuing[vii].

From the time of that realization, I could never escape a conflict between myself as a scientist and myself as a human being. Soon thereafter, I took a job at the Tokyo Metropolitan University, where I taught for about four years. After my tenure at the Tokyo Metropolitan University, I went to Germany to engage in space research at the Max Planck Institute for Nuclear Physics as a visiting fellow. Throughout those four years, however, this internal conflict never left me. Finally, I decided that I would devote the rest of my life to this issue. In the academic world, where I was a member, my colleagues refused to take me seriously. Nobody wanted to understand the issue. Since the original purpose of our research was to examine the whole history of the universe, it was considered petty to focus on the most recent decades of that history, starting from Hiroshima and Nagasaki, a very trivial timescale in comparison. My professional colleagues would dismiss my question by saying "Is he out of his mind?" The way they saw it, it was more of a problem for social sciences than for natural science. Even if I insisted that it was an important environmental issue, they would answer that it was one of waste disposal. To them, it would make no sense for a promising young researcher to study such a topic as waste disposal. In addition, there were issues such as my status in the organization and research funding, issues which could not be dealt with independently of my social concerns. It was my worry and suffering that caused me to abandon my professional position in Japan and flee to study in Germany. As I thought more about it, I finally realized that radioactive contamination had become a more and more serious conundrum, precisely because we had postponed confronting the core problem head-on.

How Scientists Think

Let me talk about the most typical reaction of lab scientists to the problem of the radioactive contamination of the environment: perhaps this summarily represents the scientific way of responding to the problem of radioactive contamination. What bothers typical lab scientists most is that it would be impossible to take accurate measurements when the environment is contaminated. It's quite common that, when you try to build an accurate measurement device, the materials of which the device is made may well be contaminated too. There are radioactive substances in our surroundings as well as cosmic radiation, so the measuring device has to be shielded with lead from these interferences. Otherwise, there will be too much background noise.

Unless you create a highly sensitive measuring device that can measure as small an amount as one count per day, you cannot really examine the history of the universe. When you want to manufacture such a highly sensitive measuring instrument, nowhere can you find clean materials to make the device since the earth's surface is totally contaminated. The most troubling among the materials is iron. Iron is used in huge quantities. It is usually purchased from iron manufacturing companies like Yahata Iron & Steel (now Nippon Steel) , Kobe Steel and JFE Steel. But the problem is that radioactive substances are attached to the walls of the furnaces where the iron is produced. Radioactive materials are not supposed to leak out of the iron furnaces, but when you take measurements with high sensitivity measuring instrument, radiation can always be detected. Human technology is not perfect. When we say "No leaks," it actually means that the level of leakage is very low.

It's the same with nuclear power plants. Supposedly there is no radiation leakage, but this only means that the level of leakage is low. There is no such thing as a nuclear power generation plant without radiation leakage. Even if there are no accidents, radiation leaks out every day. A highly sensitive measuring device will detect it right away.

Under such circumstances, scientists are in trouble because there are no clean materials available. One solution a scientist may come up with is to look for iron manufactured before 1945. This iron could not have been contaminated with radioactive substances during the manufacturing process since it was prior to the invention of the nuclear power generation technology. This sort of iron can be found in ship wrecks that sank during the Second World War. Naturally, people are engaged in the business of the salvaging and trade of iron retrieved from these ships. Such iron costs several dozen times more than the ordinary type, of course, because it is expensive to salvage and cut up the sunken ships.

As a scientist, one cannot help wanting that clean iron. The tendency is that laboratory scientists are more concerned with how to get hold of clean materials for their measurement devices than with the fact that all the surroundings are contaminated. The only concern is with how to get hold of radiation-free iron. As a matter of course, one needs a budget to purchase it. Then the next focus is on how to secure such a budget. This is how scientists typically think and behave. From their perspective, they are acting out of good intentions. Subjectively they do not intend anything evil. Objectively, however, it is far from certain that their behavior does not result in anything harmful or destructive, even though it is certain, at the least, that they harbor no evil intention. They simply want to know about the history of the universe.

Some of you in today's audience may find it strange but researchers are not capable of seeing things from a broader and an objective perspective. In my opinion, the primary factor that has driven science to the point at which it stands today is that scientists act without considering how science and technology can affect humankind as a whole; they do not take into account the relationship between science and society. At one point in my life, I myself was moving in that direction. But, as I said earlier, I felt quite uneasy about it. I could not reconcile myself with that blindness of scientific reason. I was only a research assistant then, so I rarely challenged my professor, but, once I started challenging him, we got into arguments. Later on, I began to lead a rebellious life.

Even so, it never occurred to me that I should give up my studies or leave the university. Instead, I went to Germany to give myself some time to think, but that uneasy feeling never left me. Finally, I decided to devote the main part of my study to the problem of radioactive contamination. For I believed that scientists should take responsibility for what is meant to mankind by the radioactive substances they have created. I just wanted to assert that I ought to make judgments based on my status as a human being, first, before adhering to my values as a scientist. I had no other ideological justification for my decision than this.

So I tried to work out what was important to me as a human being. I began thinking deeply about what I should do next. In the end, I decided I could no longer work for the university. After spending some time preparing, I set up the Citizens' Nuclear Information Center (CNIC) . I had always been aware of a contradiction between the law of science and the law of life. But it was only at this juncture that I came to realize that there is a clear difference between the logic that regulates science and the law that governs life. I still wasn't sure exactly what the difference was, but speaking from my lived experience I intuitively knew that a difference existed. I then felt that if I continued abiding by the logic of science, I would move further and further away from the law of life. For instance, let us recall the astonishment that I had felt when I measured radiation in places where I was in contact with people carrying out their daily activities. Only when I was with ordinary people could I realize that I had lost a sensitivity, a capacity for astonishment, about how serious radioactive contamination can be to human kind. It was frightening to think about pursuing scientific reason, and thereby becoming numb, not reacting at all, even when measuring a million counts or 10 million counts. That was the starting point of my transition. Pondering this matter thereafter, I was finally able to sort things out a little.

Man Stole Fire from Gods - Now No Life near the Fire

Now, let me talk about science and technology from a historical viewpoint. The history of science and technology can, I think, be roughly divided into three phases. In the first phase, technology was simple. Man did things "by himself," so to say, by means of his own hand, involving and risking his own body. It is not hard to imagine that there would have been all sorts of problems in this phase, too. But their impact was not too far removed from everyday life.

In this phase, science could hardly be totally alienated from the reality of everyday life. Man knew that, if anything went wrong, he would bear the consequences. So the process of trial and error worked. He was engaged in scientific inquiry by exposing his own body to potentials risks.

Then science moved into its second phase, that is, the phase of Western modernity. If I had to offer one name, it would probably be Francis Bacon. He is probably the person most responsible for the development of the procedure of positive science. According to his idea of scientific reason, man works on nature to transform and control it through experimentations. Bacon stated clearly that the ultimate goal of science is to conquer nature by the use of scientific procedure. So, in the second phase, it was believed that, rather than being in and part of nature, man stands opposed to it; man is destined to become the conqueror of nature. The conquest of nature was regarded as the ultimate mission of man.

Another idea that sprung up to support this view was the mathematical method developed by René Descartes in the 16th and 17th centuries[viii]. Mathematics is a purely logical component extracted from a scientific enterprise both complex and multifaceted. In mathematics, how things are pertinent empirically does not matter. One pursues what can logically derive from a set of hypothetical conditions, and tries to purify or formalize the logical connections between the set of presumptive conditions and the possible conclusions resulting from them. Looking at the natural sciences today, especially cutting-edge technology that is based largely on physics, scientific discoveries are expressed mostly in terms of mathematical formulae. Therefore, the availability of one or another mathematical method is of decisive importance in determining the course of science in a particular field.

In modern times scientific inquiry consists of two fundamental factors. One is mathematical method. The other is experimentation. Mathematics alone can produce only a theory, so you cannot validate the connection of a theory to empirical reality. In order for scientific knowledge to be verified with respect to empirical reality, you must conduct experimentation. This feature is called the positivity of scientific knowledge. You impose a set of hypotheses on nature and establish the validity of your theory by making nature respond to it; this procedure is generally referred to as "experimentation."

It is in the second phase that modern science and technology have advanced exponentially since the Industrial Revolution by adhering to this combination of mathematics and experimentation. Already in this phase, science was appropriated by capitalistic manufacturing technologies, which caused various negative effects, including pollution. If science and technology had stayed in this second phase, I conjecture it would have been easier for us to turn back or make some adjustments to excesses brought about by the development of science and technology.

But I am afraid that we have superseded the second phase and moved into the next one. This happened very recently. Nuclear power took us over the line. I personally believe that the use of nuclear power is beyond conventional positive science; nuclear technology belongs to a different world beyond that of modern scientific reason. This is a very complicated topic, so let me explain it in more concrete terms.

Before nuclear technology, all the achievements of science and technology were essentially imitations of nature, and technological knowledge was not alienated from life on Earth. No matter how clever we think we are, in the end, all of our scientific principles are derived from nature. One example is the airplane, which is a consequence of our efforts to imitate what a bird can do. Even with all of today's impressive and powerful aeronautic technology, we still do not have the skills to build an airplane that can take off practically anywhere without a runway and change direction with minimal energy, like a bird. What is even more impressive than a bird is a creature like a mosquito. I wonder where in that tiny body all the energy needed for flight is hiding. Man will never be able to imitate that mechanism. An example of a very clumsy imitation is the jumbo jet, which cannot fly without massive amounts of fuel.

Essentially all these technological achievements that we enjoy today are based upon the imitation of nature.

In some sense nuclear power is also an attempt to imitate nature. But there is one crucial difference. Among Western mythological tales, we know the story of Prometheus, who stole fire from the Sun. This act enraged Zeus, who punished Prometheus for his transgression. This story is particularly symbolic as far as I can see.

Prometheus stole the gods' fire. Likewise, I see nuclear power as a kind of stolen fire from the heavens. It's certainly not earthly fire. Stars shine because of the nuclear reactions going on in their core. Our own sun shines because it is burning hydrogen. This is the same mechanism that gives a hydrogen bomb its power. In short, it's a sort of nuclear reaction. If you get close to the sun, you would be exposed to massive amounts of radiation. Of course, you would die just from the heat, but even if that weren't an issue, the amount of radiation coming off the sun would prevent anyone from getting near it.

So the point is that there is absolutely no life on stars, not even near them. Man's technological prowess has advanced considerably, and we can now see things at an extremely great distance. With today's technology we can find out about stars billions of light years away. And yet, we have found not a single star that supports life. So the probability of discovering stars with life is very low. There might be one such star but my guess would be that it's probably far beyond the reaches of our communications. And if there were one hundreds of millions of light years away, what good would that do us? Suppose we send a signal. It takes hundreds of millions of years to get there and the same amount of time for us to receive the response. In total, it would take nearly a billion years. So, if we send a signal now, people a billion years from now will receive the response. I don't think by then there will even be any humans on Earth because of the reckless way in which we are acting now. So, I don't think there is any other life in the universe.

If this is true, it means the Earth exists under very special conditions, one of which is the fact that the Earth is protected from radiation. Another factor is probably water. Water is also involved in the blocking of radiation. In any case, what is significant is that the Earth is insulated from radiation.

I will not get into the mechanics of this here, but it involves things like atmosphere, the magnetic field and our distance from the Sun. Suffice it to say that the Earth is protected from radiation.

When the Earth was formed, radiation was extremely strong. So there was no way life could exist. It's said that the Earth was shaped into its current form about 4.6 billion years ago. Although this remains somewhat arguable (as I mentioned earlier, at one point in my life I studied the history of the universe so I did look at a lot of the relevant data). While the Earth is thought to have formed about 4.6 billion years ago, the early solar system was shaped 5 billion years ago. Back then, radiation was much, much stronger. It is believed that star particles coalesced to form first the solar system and then the planet Earth. These particles contained lots of radioactive substances so they were very hot, in the radioactive sense. It took 4.6 billion years to cool down. When the radiation was reduced to a sufficiently low level, life came into being. And then, just when all the conditions for life to exist on Earth were finally satisfied, man came along and stole the gods' fire. He split the atom. He created—by his own deliberate undertaking—nuclear power and radioactivity. Human kind did something extraordinarily contradictory and excessive to the conditions of life on Earth.

In the myth, Zeus punished the man who stole the gods' fire. Let's not forget about that, because nuclear power is indeed the fire of the Heaven, or at least the fire of the stars. It could not have been created here on Earth. The instant man stepped into this new field, science and technology shifted to a new phase. Until then, man had been trying to imitate nature on Earth. But with nuclear power, man began to imitate what is inherently of the Heaven.

Life on Earth has its own governing principles. There are laws that govern the creation of life on Earth and laws that govern the continued existence of life on Earth. These are the focus of my lecture today. However, the intellectual development of mankind has made it possible for us to access a power completely at odds with these laws of life on Earth. This in turn led to the tragedies of Hiroshima and Nagasaki as well as to other issues related to nuclear power that have been troubling us ever since.

Is it with Life or against it?

Taking into consideration the issues of environmental destruction and radioactive contamination brought about by the development of science and technology, we probably need to be clear that some forms of science and technology do accord with the principles of life on Earth while others do not. Nuclear power is a particular case, but there are many other forms of science and technology that do not accord with the principles of life on Earth. It is not only nuclear power, but more generally that we cannot necessarily reconcile the law of life on Earth with the law of science — this is the topic for the latter half of today's discussion. In the field of biotechnology, humans have now mastered some skills with which to modify genes to a considerable extent. There have also been great advances in medicine, making it possible to transplant organs from one body—even a dead body—into another, and with good success. Regarding death itself, a person is declared dead following brain death, which is also a factor in organ transplants. These illustrate the kinds of things happening in today's world. As far as I can see, these ideas are clearly not based on the conventional definitions of a "natural death" and a "natural life."

This is a very difficult topic, and one best suited to the latter half of today's discussion. It's something that we must all think about together. Within the scope of the traditions of Western science and technology—the field in which I was trained and pursued an academic career—a human being is basically defined as an animal that thinks. Ultimately a human being can be reduced to his brain. In fact, science and technology is a product solely of man's cerebral functions. That's why it doesn't deal with mankind as a whole, or man as a natural being living in environments of many sorts. This is what gives rise to the idea that whether or not the brain is functioning determines whether a person is alive or dead.

A question frequently asked within my own academic circles is this: Now that it has become possible to transplant organs, will it be possible in the future to replace a malfunctioning hand with a healthy one? Well, this is very likely. It may even become possible to replace limbs and organs all at the same time. If this comes to pass, we will be confronted with a new question: If all of the body parts have been replaced, is it still the same person? A typical response by the Western way of thinking to the question is that you can replace any other part but the brain and the person remains the same person, as long as the original brain remains intact. But is this right? I'm not convinced.

Suppose that the brain, which is only a small part of a whole human body, is the only original part left of Person A. All the other parts and organs used to belong to Person B. Where then do we draw the line? How can we say that the person is Person A? This is getting into the realm of science fiction, but the people who work in this field say that we are coming very close to being able to replace organs in this way. So technology has already taken us beyond what we thought was possible. And we are barely keeping up. Most of us aren't even aware of what's really going on.

In the second half of my discussion, I'd like to set out the general issues of the law of life by examining the issues surrounding nuclear power, and also propose possible directions in which we may be able to solve these problems.

PART II

Earlier, I mentioned that the gap between the logic of science and the law of life is getting wider and wider. I want to return to that point just to clarify the context of my argument.

First, I will talk about it from the perspective of natural sciences. I would like to show how such a gap is conceived of in terms of natural scientific vocabulary. First, this much can be ascertained: life on Earth as we know it is dependent on nuclear stability[ix]. In other words, life on Earth constitutes the world of chemical reactions. From a scientific perspective, human bodily functions are basically chemical reactions, though this may be rather simplistic to say. The human body produces energy by burning or oxidizing various substances. This is all done through chemical reactions. Take, for example, a gene. A gene is basically a chemical substance called DNA ruled by the principles of chemical reactions, which have nothing to do with nuclear reaction. Chemical reactions are concerned with what goes on outside the atom and not with the internal dynamics of the atom. This state of clear differentiation is what we mean by nuclear stability.

The stability of the nucleus itself is fundamental to life. To clarify, it could also be called "atomic stability[x]."

Nuclear power, on the other hand, relies upon a technology that generates energy by destabilizing nuclei. If the nuclei are not destabilized, there is no nuclear power. This simple fact lies at the heart of all problems related to nuclear power. Secondary problems like the malfunctioning of valves may arise but they are not the key problem. Technologically speaking, there are many problems that could emerge, including the mismanagement of control rods. But here I'm talking about the fundamental issue, which is not something that can be solved by improving the technology. I'm not saying "no" to nuclear power because an adequate technology for its safety has yet to be developed. My opposition to nuclear power is not based upon the historical immaturity of its technology. My point, and what is of greatest importance, is that by its very nature nuclear power is in conflict with the very law of life.

The Inextinguishable Fire

This is why I often tell people that nuclear fire cannot be extinguished. Sure, it is possible to halt the operation of a nuclear power plant by performing various technological actions, including operations concerning the fuel rods. But even if you stop the plant, that doesn't mean the nuclear fire has burned out.

This is why the problem is a grave one. Radioactive products accumulate inside the control rods and keep producing heat even when the plant is not in operation. With the Mihama accident[xi], the nuclear reactor stopped operating at 13:40:39 hours but this didn't mean there was no longer any danger. In fact it was the subsequent leakage of primary coolant that necessitated the activation of the Emergency Core Cooling System.

This was an emergency quench measure. But the point is that such a measure had to be taken because the reactor core was still very hot. And why was it still hot? Because the radioactive products, sometimes called “death ashes,” inside the rods kept generating heat, even after the reactor stopped. The fire had not completely burned out[xii]. This is the fundamental problem. Nuclear material cools down very slowly. It takes an incredibly long time. As I always say, the impression the phrase “death ashes” gives is half right and half wrong. It is certainly right in the sense that the ashes are horrifying since they are lethal. But death ashes are not really ashes. Ashes are generally thought to be cold, but death ashes are not. Even if the coolant leaks out, heating continues even after the reactor stops, and this will conclude in a nuclear meltdown. This shows that the calorific value of death ashes is in fact extremely high. So I tell people that they are not actually ashes, but instead embers.

The fire stops blazing but the embers continue to glow. I think this is a good metaphor to explain what actually happens when a nuclear power reactor stops operating. Some of you might have seen several documentary films about the 1986 Chernobyl nuclear accident. If you have, you may remember seeing a shot of the glowing reactor core. The cameraman who filmed it must have put his own life at risk to do so. The reactor core glowed because it was more like embers than ashes. Substantial damage was done to the facilities, and the reactor stopped, and yet the core was glowing. That's the ember. It is the most typical state in which you find radioactive ashes after the reactor is stopped. Now, let me explain how the embers cool down. Radioactive substances have what is called a "half-life[xiii]." Some substances have a half-life of tens of thousands or even millions of years, meaning that the radioactivity will not disappear for millions of years. So, a radioactive product is not something that man can extinguish, even if he wants to. It cannot be extinguished with water or chemical extinguishing agents or neutralizers. It's an inextinguishable fire.

I myself participated in the early stage of nuclear power development. We scientists were intent on trying to create something bigger and more powerful in the shortest period of time. And we made considerable progress in building a large nuclear power plant and burning the atomic fire. There's no doubt about that. The new plant in Kashiwazaki-Kariwa[xiv] has a huge 1.356 GW nuclear reactor. We have become skilled at keeping the fire burning, but, not at putting it out. We have made no progress at all in that regard. So however far the technology has advanced, it is a one-sided progress. That's why we cannot say that man has control over it.

In order for man to master certain technologies and gain control over energy, he must be able to switch it on and off at whim. But atomic fire cannot be extinguished so freely, as we can see from the fact that reactors will continue to emit radiation even after they are shut down. Specifically, what's going on is that excess energy is being emitted in the form of radiation. This is because the fire has not completely burned out. Think of it as a kind of heat. It's not gone completely. It's still smoldering. After the Chernobyl disaster, radioactive substances like cesium ended up all over the world, heavily contaminating food. To say that food is contaminated means that the ember is still burning inside the contaminated food. When that food is taken into the human body, so is the ember. Then, so to say, radioactive substances continue to burn inside the human body. There is no way that this can be good for human bodies.

Not knowing what to do with this smoldering fire, government officials, corporate executives of nuclear industry, and nuclear technology experts are now talking about dumping it in the village of Rokkasho in Aomori Prefecture. I think that this plan to build nuclear fuel cycle facilities including a reprocessing plant at Rokkasho best exemplifies the irresponsibility of our society. In fact, I have published quite a long book about this subject, and am determined to devote the rest of my life to putting a halt to this plan. Here is a copy of that book. It is a compilation of my work over the last several years[xv].

My decision to publish this book was shaped by my view of the mentality of scientists that I described earlier. As you know, I used to be engaged in the production of radioactive substances. Seeing how thoughtlessly nuclear waste is handled makes me feel guilty. I feel the same about seeing how the outcome of that work is being thrust upon the people of Rokkasho, who have decided to accept the nuclear recycle plant because their village is suffering economically due to rural flight.

Driven by this sense of culpability, I'm searching for a way to stop this plan from being carried out. But, even if I can stop the construction of a recycle plant in Rokkasho, I would be able to suggest no other place as an alternative. There is no such thing as a suitable final destination for radioactive substances. Unless I repent for what I have done by stopping this reckless plan, I cannot help but feel that I haven't taken full responsibility for choosing nuclear chemistry as my career. The contradiction between the law of life and the logic of science is so fundamental that, even with such strong determination, I still cannot find a solution to the conundrum inherent in the inextinguishable fire.

By criticizing the Rokkasho project, I wanted to show, to myself and others, how inexorably different the law of life that demands human beings to act in everyday life is from the laws of science that apply to the world of nuclear science and technology.

The second very serious problem is that the speed at which things happen in the nuclear world and man's sense of speed are totally different. The pace of change in the nuclear world is absolutely different from the speed of man's judgment or response.

Man cannot endure this gap in scale between the nuclear and the human worlds in temporality. In the nuclear world, one tenth of a second, or even less, can make all the difference. Assessments must be made in real-time. Even a second too late, and the result could be disastrous. But man cannot live like that. However advanced his technology is, man can never react fast enough to keep up with the nuclear world. Cinema provides us with a good illustration. I think you know how cinema works. Still images or "frames," each slightly different from the one before, are projected onto a screen in very fast succession. This produces the illusion of movement. It's a trick, a visual illusion. If our eyes were faster, we would be able to discern the individual frames. The trick only works because our reaction time is slower than the projection speed. This is fine for movie. However, it's not fine when we get into the realm of nuclear power.

During the Mihama incident[xvi], numerous signs appear to have shown up an hour before the accident took place. When the charts were analyzed afterwards, they showed that the level of radiation had risen gradually. But while it was actually happening, the central control room couldn't discern a problem. That's why the accident occurred. Sure, we can blame the operators by saying that they are so accustomed to having lots of different things going on that they may not have noticed an incremental change on some meter. But we must recognize an aspect to human capacity, because of which it is very difficult for a human agent to make an appropriate judgment, even in ten or twenty minutes when changes are gradual. Consequently blaming the operators misses the point when it comes to determining the cause of the accident. Human capacity has limits. We should try and be careful not to overestimate the capabilities of individual operators. If the value jumps by a factor of 100 in an instant, it will catch the operator's eye. But if it rises slowly by 10%, 20%,

30%, it will only cause him to ponder the situation over a longer time span. If it is his first experience with something like this, he will tend to be very cautious and will probably take even longer to make a decision and act.

So, what must be acknowledged, rather than blaming the operators, is the absurdity of putting a man with limited capabilities in a situation where he is required to possess unlimited capabilities.

This brings us to the third problem. Similar to the second problem of speed, there is the issue of human error, which is simply unacceptable when dealing with nuclear power. Today, we live in an era when dramatic accidents can have major impacts. As I wrote in my book[xvii], The Era of Catastrophic Disaster, when it comes to the huge systems we have today, we can't afford any errors at all. We have things like failsafe and foolproof systems, which cause a machine to stop if a minor error is detected. In addition, it is generally assumed that machines compensate and correct for human error, while safety inspections are conducted with the presumption that failsafe and foolproof have been included in the huge systems by design.

But if we look at actual accidents that have occurred, we find that there are a number of cases where the failsafe system actually failed to work. The failsafe system cannot respond when humans act in a manner that is entirely unpredictable for the system. So failsafe and foolproof systems are effective only for a limited range of accidents.

In *The Era of Catastrophic Disaster*, I analyzed the various types of accidents. The one that most interests me now is the domino reaction. Suppose there is a small incident, like a valve malfunction or a very small perforation. Normally the operators are stationed in the central control room, but unless something serious actually happens there is virtually nothing for them to do. For them it's a walk in the park. I've heard that some even play cards. In the US, there was actually a case where an operator was dozing off when an accident occurred. He had to pay a fine later on. These things really happen.

If nothing goes wrong, the machines function automatically, making the operators extremely bored. Every single day, the same old thing! Then, all of a sudden, comes something extraordinary. The operators panic and can not respond to the situation. Normally, their job is easier than driving a car. In fact, driving requires more concentration. Nuclear power plants usually operate automatically. But once something happens, it becomes the operators' responsibility to deal with it. In their confusion, they make an error, like pressing button Y instead of button X. The machine reacts accordingly, doing something the operator did not intend. This just makes matters worse for the operator. And a series of negative chain reactions can easily ensue. Thus, something minor develops into something major.

A very good example of this is the Three Mile Island accident in the United States. All these little things happened that were insignificant in and of themselves. We have learned many lessons from this accident. But its overall cause was, perhaps unjustifiably, attributed to undertrained operators and miscommunication between man and machine, rather than to machine failure. As a result, the control room was redesigned and the location of the buttons was changed. But such a remedy cannot solve the underlying problem.

In an extreme case, even the operator's mood can have a big impact on things. That is why psychiatrists and psychologists are stationed at nuclear power plants in the United States, to monitor the mental state of the operators.

Japan has not adopted this habit yet. We just do a background check on job applicants before employing them, and maybe some kind of ideological screening, too. The United States is not so thorough in their checks, so operators often cause problems by using drugs in the control room. That's why they need the psychiatrists.

Whatever the case, the point is this: there is no room for error. This is actually pretty worrying because to achieve a completely error-free environment we would essentially have to transform man into a machine. In other words, rather than bridging the gap between machine and man by making machines more human-friendly, we would instead be bringing man closer to machine. *In order to conceal the contradiction between the logic of machine and the logic of human life, we try to transform human beings into machines.* We must understand that we are living in a world in which such a perverted remedy has been adopted.

Anyway, things do not always work out like this, and so accidents do occur. With respect to Chernobyl, the causes have been defined as design error, human error, and so on. Whatever the case, wherever human beings are involved, accidents are inevitable. Incidentally the same goes for design error. As long as human beings create the design, things like human carelessness and oversight are unavoidable. Now, you can see why I find it really frightening that, wittingly or unwittingly, we humans are forced to enter a world where there is no margin for error.

I know from my own experience that to err is very human. When I worked in the nuclear industry I never caused a major accident, but I did cause a few small ones, and the lab became contaminated quite often. I doubt that there is any expert handling radioactive substances who has never caused some kind of contamination. One thing that happens quite often is that you are working with radioactive substances in a container and all of a sudden you realize they have disappeared. In the lab where I used to work, small explosions and uranium contamination of the whole room were not rare at all.

Man cannot avoid these kinds of problems. Even experts cannot avoid making mistakes. Technology that makes some allowance for mistakes, therefore, is more acceptable. Technology has grown beyond the scale of experimentation capacity, and technology that doesn't allow for mistakes is just horrendous.

There is the idiom "trial and error," which leads to the fourth problem. Today, nuclear power technology has become too large in scale for us to verify it through experimentation. Of significance is the fact that nuclear power technology is too large to experiment with. As I said in the first half of today's discussion, great progress was made in science during its second phase through the procedure of experimentation. It is through trial and error that science and technology have progressed. But now, in its third phase, the age of experimentation is over. We have entered the phase of scientific reason that no longer allows us to appeal to the procedure of experimentation.

During the recent accident at Mihama, the Emergency Core Cooling System (ECCS) was activated. In a way, this was an experimentation of the ECCS. It went well this time. I'm not sure that we can say it was an unmitigated success and that no damage was done to the reactor. However it did work. I might also point out that in the entire world there have been only a dozen or so cases where the ECCS was activated. In one case, it didn't work out so well and the result was what we now know as the Three Mile Island incident. At Chernobyl, the ECCS couldn't be activated from the start. This is to say that it is not possible to conduct a meaningful experiment or drills, and there are only a few actual examples that demonstrate how the ECCS works when activated.

Today what is done instead is to isolate two to four fuel assemblies, and an electric heater is used to simulate the reactor core of a commercial nuclear reactor, which normally consists of 700 to 800 fuel assemblies. This simulation does not actually induce nuclear fission. All it does is heat the few assemblies, to pour in water, to let the tube rupture, and to activate the ECCS to see if it functions properly. But, because this represents a scaling down by a factor of several hundred, we have absolutely no idea whether this small-scale model can really be called a simulation. Water flow is also very different. And even when we are asked whether the trials on this reduced scale are producing satisfactory results, we can only answer, "not really." Things just do not go the way we expect. Experimental results do not quite match the computations, and the cooling down does not go as calculated. When the ECCS is activated after the tube rupture, water from the ECCS sometimes leaks out before it manages to cool down the core. In a sense the real trial or drill is conducted when an accident occurs at an actual commercially-operated reactor. Again, this does not produce perfect results, as we can see from the Three Mile Island accident.

Subsequent to that accident, it was found that computational models failed to determine what had actually happened inside the reactor. The damage was not as extensive as Chernobyl, so all the charts were available. All the basic data, like changes in pressure, could be gathered from the control room, although some of the thermometers were broken. An initial attempt to use those data to model the accident on the computer failed. It took three months of round-the-clock operation of a large computer before the incident could be successfully replicated. The general conclusion from that model was that the core had not melted.

But the story doesn't end there. When the reactor was finally opened after several years, 70% of the core was actually melted. This is the best that the human intellect could possibly predict; this is where the grand sum of our science and technology gets us. Herein lies man's conceit: the belief that science and technology are all-powerful. Indeed it is frightening that we are now deprived of the means of experimentation.

Nevertheless, man will continue to advance much further. If there is nothing to stop him, the human intellect will make great strides. But that does not mean that we will overcome our weaknesses. On the contrary, our weaknesses will be even more exposed. Any achievement made will be one-sided. That is, man may learn more about lighting the fire but will not use his intellect to consider what impact that fire may have on himself or society. He won't evaluate the potential harm and devise any meaningful protections.

This has become a threat not only to mankind, but to the natural world. The Gulf War[xviii] clearly reconfirms my position. People debate whether Saddam Hussein or George Bush is at fault, but that's only relevant within the realm of human society. It doesn't matter to nonhuman creatures living in the desert or the Persian Gulf whether it's Hussein or Bush. All they desire is for man to put a stop to this whole senseless event. From now on, people must start to adopt a similar viewpoint. The Gulf War has shown us two things: that the world can be seen from non-human perspectives, and that advanced technology can be very destructive to other non-human lives. The Gulf War makes me think that technological advancement has brought mankind to a point where it cannot afford wars. The level of damage left by the Gulf War is devastating proof of this.

While science and technology can now have an impact on the entire globe, man, who is wielding it, is not aware of this at all. He is making weapons just to see how far he can push the new military technology, the Tomahawk, the Patriot, the Scud missile, depleted uranium bombs. I'm very disappointed and disturbed by all of this.

Earlier, I talked briefly about the speed of radioactivity. Now I would like to touch on the subject of how long radiation lasts. In fact, the length varies. Some substances will stay radioactive for tens of thousands of years, some for hundreds of thousands of years, some for millions of years. The longest is two billion years. But of all the types, the most troublesome and toxic of radioactive substances is neptunium-237, a uranium by-product produced inside nuclear reactors. It has a half-life of about 2.1 million years, which means it takes that duration for its radioactivity to halve. It is also very toxic.

Burning one pellet generates enough heat to produce one year's worth of electricity for one household. It is said that this is the great benefit of nuclear power. But, from the viewpoint of life or humanity, it must be considered in terms of the negative impact its nuclear waste can have on human life, rather than any positive benefit. The nuclear products contained inside one pellet are equivalent to 50,000 lethal doses. From the viewpoint of waste product disposal, what we are doing is simply nonsensical, not to mention extremely hazardous, just for the sake of one year of electricity for one household.

It goes without saying that 50,000 people are not going to die immediately from one year's worth of electricity for one household, but we are producing something with a lethal capability. Moreover, neptunium-237 is a component of the nuclear cycle. And with a half-life of 2.1 million years, it's going to remain on Earth for a very long time. In one million years, the toxicity of the substances left in the pellet will be reduced almost to the level of potassium cyanide. Potassium cyanide is a chemical poison, or Earth's poison. It takes approximately a million years for a poison of gods' fire to be reduced to the Earth's poison. In other words, we are producing in the world of human life a heavenly poison that requires an infinitely long time to become an earthly matter.

So far I have mentioned five basic problems about nuclear power technology: first, nuclear instability, second, the speed required in the nuclear world, third, the impossibility of conducting experimentation, fourth, the fact that there is no room for error and, fifth, the infinite time required for radioactive substances to become harmless. Each of these represents a contradiction between the law of ordinary life and the logic of nuclear power.

If they were matters pertaining to another world, then I wouldn't mind. But nuclear power plants exist as real things, standing alongside all the other things in our daily lives. And what's happening inside the reactor cores is totally a matter of nuclear fission, in the gods' realm, so to say. Inside the buildings that house nuclear power plants are people who actually engage in their operation and control. Further, the steam turbine is a classical machine. So the nuclear world is not independent of things of our world, it is not completely self-contained. Things could be better if the logic of the nuclear world did not affect us, but unfortunately it does. And there are people, like marginalized migrant workers, who mop the floors inside the nuclear plants. Clearly these nuclear power plants are not entirely modern, state-of-the-art facilities. Moreover, there are ordinary people living just outside these plants.

The world of nuclear power is not independent of the rest of the world. Human beings live according to a different law of life and to a different speed of life from the logic of science and the speed of the world of nuclear power. The timescale of a man's life is totally at odds with the timescale of the nuclear world. And this contradiction is hardest to deal with. Having been involved in nuclear technology, this contradiction is the hardest to stomach. In fact, this has been my driving force, rather than my fear of radioactivity. What made me decide to sever ties with the nuclear industry was that I could not make peace with its technology.

So now on what am I going to do? I have no clear plan but I will keep speaking out against nuclear power plants and do whatever I can to prevent the inept disposal of nuclear waste. As a person who has been involved in the field of nuclear chemistry, I must take responsibility. The question is how. To work as an expert at a university or in the private sector, would instantly blinker me. Instead, what I want to do is to go down to where people spend their lives. "Going down" may not be an appropriate expression, but what I mean that I want to live among ordinary people and analyze the nuclear problem by drawing upon my specialized knowledge on the opposite side to the logic of science and technology. I must rely upon my scientific and technological expertise, but I stand on the side of the law of life and speak first as a human being. The challenge will be to see how far I can go and what effect I can bring about.

Limits of Scientificism and a Return to the Law of Human Life: Imposed Rationality - Conclusion

Having summoned the courage to leave the university system, I became a citizen scientist. This has been my role now for the past 15 years. It may sound as if my decision was based on my career choices as a nuclear expert, but it had nothing to do with my professional career. I believe each of you faces the same problem, though maybe in different ways. This stems from what one might call the logical structure of today's society, according to which large cities continue to grow while the flight of people and capital from the countryside is unstoppable. Unfortunately such depopulated areas are ideal sites for nuclear power plants.

The rejuvenation of depopulated country areas requires money. Hence local governments and private industry can justify pumping huge investments into impoverished rural communities; these investment policies mobilize the logic of science and technology rather than the law of life. How much can I resist this trend of taking us away from the law of life when I go back to basics as a man? This is the question every one of us faces. I cannot say anything profound about the nature of mankind because I'm only a scientist, but if I speak from experience, I think man is a very ambiguous creature. He cannot rationalize everything or quantify things in terms of numerical value. I use the phrase "imposed rationality," by which I mean that man cannot do away with rationality regardless of whether he likes it or not. When man is forced to be involved in nuclear power, he just rationalizes what he is expected to do, but when he decides to go against nuclear power, he also has to have professional expertise to be able to make his point rationally. Whether at a public hearing or on other occasions, he will be powerless unless he wins the argument with logic.

It should not be like that. If there's something you don't like, you should be able to say so out of honesty. But that is easier said than done. Especially people like me must show my professional expertise in order to persuade the public and authorities; on certain occasions, we must use mathematical formulas to counter-argue the justification proffered by corporations and the government in favor of nuclear power plants. Imposed rationality, in other words, is the state in which one is forced to have a rationalist perspective and justify one's life by rationalism. I think imposed rationality has a very strong effect in every sphere of life.

• Errors Are Becoming Increasingly Unacceptable

Another concern is that errors are becoming less tolerated. Today, education systems are living by this mantra. Instead of aiming at developing individuality, they focus on nurturing conformist individuals so as to reduce errors. This inculcates stereotypes into students and controls their thoughts, but, as I said earlier, science and technology is potentially largely responsible for this tendency. Perhaps companies demand such people, and so education systems have followed suit.

• Finitude of Human Existence: Human Intellect Cannot Predict Our Future

We are living in an age where we must act without knowing what effect our actions will have on the future. We cannot see the future. Our puny human intellect does not allow us to predict what the world will be like in 10 or 20 years. As we approach the turn of the century, it's becoming quite popular to make predictions about the next 100 years. I myself receive such requests, but I usually decline since I simply cannot forecast the future. The human intellect cannot see that far ahead when it comes to science and technology. We can't even predict what will happen tomorrow, really. Man is not that kind of creature. This makes it sound like man is useless. However, I don't really think so. Man is connected to the wider world through some means other than scientific logic. I believe he can use his imagination to consider how to move forward. Through life, he is connected to something beyond our existence.

• A Responsibility toward the Future

A movement called Hansalim led by Kim Chi-Ha is becoming popular in Korea nowadays. It is probably based on the same ideas as Donghak. I know Mr. Kim personally, and thus this movement has had an impact on my life. His idea is that the universe is the human being and vice versa; that is, as one existence in the universe, a human being is living his own life this very instant. While your life is one as an individual, you are also living your own life as a creature that exists for a very brief time in the long history of the universe, and there are other individuals before you and will be after you. As Kim says, "The life of an individual begins at its birth and ends at its death, but I'm not living my life only as an individual. Rather, I embrace the whole universe inside me." His ideas are a little too metaphysical for my taste, but I think they point to the essence of mankind. I also believe that a life does not belong only to one person, but to those who came before and those who will come after.

I feel a little silly saying this in front of you, as people who think about life and death every day, much more than I do. But I hope that as someone who has studied the natural sciences, I may have something worthwhile to say.

• From a Science of Individualism to a Science of Cohabitation

Until now, science has been all about standing out. By "standing out," I mean being stronger, faster, larger than the competition. However, that has set mankind apart from the natural world. We now suffer isolation from other species. At the same time, within human society, there is now a gap between those who benefit from science and technology and those who don't. I've insisted on prioritizing the laws of life over this kind of science. Unless we all start to give this idea serious consideration, our future will be doomed.

It's too optimistic to take for granted that man is capable of controlling science because he created it. At the same time, there is much room inside science—I mean science subordinate to man—to transform itself. There are slow moves being made away from the science of "standing out" and toward the science of cohabitation, a science that coexists with life. I detect a beginning, an intimation of a transformation of science

I have a clear vision about what kind of technology I would like to see develop. But, as I mentioned earlier, what's most important is for scientists to start thinking about science from a different perspective, rather than just thinking about their own research, daily tasks and problems. This is something that I always keep in mind.

• A Voice for the Voiceless

One of the things that drive me is the sense that I have a responsibility to the dead; I am responsible to make the voices of the dead heard through my words. These days I feel like I'm living alongside the dead. Having been a part of the nuclear industry, I get the sense that the spirits of the victims of Hiroshima and Nagasaki have not yet found peace and that they are still raising their voices.

I am concerned then with working out what they might be saying and how to express it myself. If I may refer to more recent events, we keep seeing that photo of a cormorant covered with oil in the Gulf War. That bird cannot speak out in its own defense, so I believe it's our responsibility to speak for it. The extent to which we are successful at speaking out on behalf of the deceased and the nonhuman creatures with no voice of their own will determine the extent to which we can pass on a better world for future generations.

We are accruing heavy burdens for future generations. Another question is how we can speak out on their behalf. This is a common sharing beyond generational boundaries: a common sharing with the dead of the past and with the yet unborn.

The idea of coexisting with nature has recently become a hot topic. I have advocated for it myself. In the context of science, we are being challenged to see how far we can go. There is a limit to what I can achieve but I will persist. I don't know to what extent I have managed to address universal and general topics, but I believe I have expressed how I would like to think and live.

Please accept my apologies for exceeding the allotted time. Thank you for your attention.

Q&A Session

Q1: We have already accepted the presence of nuclear power plants. If we are to now oppose the transfer of the waste these plants produce to Rokkasho, then where are we going to put the spent fuel rods?

A: To be blunt, the plant that produces the waste should take care of it. This, I think, is a good general rule. In my opinion, when you accept the plant, you have to understand that you are also accepting waste. There are just no better alternatives. It's a bit inconsistent to accept the plant but refuse the spent fuel rods. Transferring the rods from one place to another is difficult and therefore very risky. One commentator has said that nuclear waste should be allocated to each prefecture in accordance with its electricity consumption. Although this seems like the morally fair option, it's not really acceptable, in my opinion.

Q2: Was the fact that atomic fire cannot be extinguished already known at the time of privatization of nuclear power plants?

A: No. I think people involved in the initial planning were very optimistic about the whole thing. First, nuclear power technology sprang out of the development of military technology. These people were not in the habit of thinking about what might happen in the future, nor did they bother to. Because the technology was designed to hurt people, safety was almost irrelevant. Then, at the time when we began to get involved, the plan was for "the peaceful use of nuclear power," so safety was taken into consideration to some extent. Yet it was thought that technology would eventually solve the problem, and that toxicity could eventually be eliminated. But all subsequent attempts to do this failed. Since the history of science leading to the discovery of nuclear power had been a long succession of surprising discoveries, it had been expected that, if that trend continued, the technology would advance to the level where a radioactive substance, once created, could be transformed into a nonradioactive one. But it's almost certain today that there's no way to do that.

Q3: Do you have any specific idea of what it would be like if the science and technology of cohabitation were to be actualized?

A: The recent trend is to try and blur the boundaries and take a more nuanced view of man. But I don't buy that. If you look at it just from the technological viewpoint, I think you miss the essential point.

It is important that science and technology not be in the hands of a limited cadre of experts but instead should be open to everyone. Some stories by Kenji Miyazawa provide an image for the science and technology of cohabitation, but it has not been pursued[xix].

At the Citizens' Nuclear Information Center (CNIC)[xx], which I run, we do not believe it is necessary to have experts. Instead, we understand that there are many things that ordinary people can learn and do. We have had some modest success. Ms. Mikiko Watanabe, the co-author of my book, Nuclear Fallout on the Dining Table[xxi], which you bought today, is not an expert. As a mother, she became deeply interested in food contamination and was determined to do something constructive in that area. That was the starting point. I want to respect individuals who demonstrate this kind of highly tuned awareness and ambition. Ms. Watanabe is now one of the most knowledgeable people in Japan when it comes to food contamination. She does not consider herself an expert though. I, for one, think that there is a possibility that if we can engage in science with that kind of awareness we can give rise to a new type of science non-contradictory to the law of life.

At the Citizen's Nuclear Information Center (CNIC), there are equal numbers of male and female staff. There has long been a general perception that science is done by men. In fact, if you go to scientific academic conferences even today, you will notice that about 95% of the participants are male. This stark gender divide is very rare in contemporary society and what comes out of this male-dominated world is quite troubling. I want to challenge this trend by creating an organization that better reflects everyday experience. That's one thing. Even in the scientific sphere, consideration is increasingly given to the environment, nature and the passing of time. But what is important for the time being is to train people who can assess the negative effects of scientific development with the same gravity as those who are involved in scientific development. There is not a single school at any Japanese university devoted to the assessment of the negative effects of scientific development, although there are plenty of people on the development side. I do not deny the positive aspects of science but the other side of the coin cannot be overlooked. Scientists only deal with the positive aspects. Research into the negative aspects hasn't been developed and there are no people to do it. The Citizen's Nuclear Information Center has yet to make a lot of progress, but more than anything else, it's important to develop this idea further.

Q4: Are there any moves among the scientists and engineers involved in nuclear power to respect the sensibilities of ordinary life, like you say?

A: Yes. A lot more than there used to be. When I first advocated the idea, there was no one to support me. Now there are people who support the idea, and quite a few of them are affiliated with national research institutes. They do not all think in exactly the same way. Many of them have their own interpretations. And there are quite a few people at nuclear power companies who agree with us. The thing is, it is difficult to speak out or take action while working for a Japanese company. Some of these individuals provide us with data and discretely cooperate with us, but because they rarely speak out, unfortunately it is often not clear to us what they really think. If I were in their position, I would care less for superiors and colleagues in the company and would do a lot more. But, of course, I cannot oblige them to do the same. Certainly there is a risk of being fired. I am tempted to say life is more important than one's career. But anyway, I get phone calls from these people, especially when a nuclear accident occurs. In fact, I got quite a few calls after the Chernobyl incident. There are many people I used to work with who tell me that they are frustrated. But that does not mean they have decided to quit. Only a few have. I think those with strong convictions account for less than half, maybe 20% to 30%.

Q5: Science cannot answer every question. How do you account for scientists' beliefs? That is, knowing what they know, how can they possibly continue business as usual?

A: They may think that, given enough time, the problems will solve themselves. If you stay locked up in your lab all the time, you lose your feeling for life and do not ever think about the risk of radiation. They believe in scientificism. Twenty percent of them think that technology will eventually solve any problems. Fifty percent have given up. They think that even if they do act, it will not change anything. "The nuclear power structure is too huge to fight against," "the political party in power is too strong," "the world won't change," "I don't have the money," "I have a family to take care of," "I might as well give in." This is no joke. A relatively large number of people think no matter what they say, nothing will change, so they keep their mouths shut. The number of people who think nuclear power plants are good is very small, but this doesn't mean anything. Instead, they settle for subsidies. I think this is one way of giving in. In fact, this feeling of resignation dominates today's world.

If people held out hope for the future, they would be more active and aggressive. As far as I know, no one at electric power companies or nuclear power plants is satisfied that they are doing a good job of contributing to their colleagues, local communities, society and humanity.

They all long to retire. They would be pleased to be asked to resign so they could be finally free of nuclear power. In other words, they are afraid that a major accident could happen on their watch; it would surely cause a huge amount of trouble. The pressure on them is tremendous. Some people have quit just because of that. As you can imagine, because of this despair, they are not actively engaged in their work. Their pessimism, the fact that they are engaged in the nuclear power industry because they have no other choice, is kind of scary. If these people were put in the control room, they would tend to act recklessly and be highly likely to cause an accident. This is an issue that runs through the whole nuclear power industry. The Rokkasho Project is also quite a sloppy plan, reflective of this pessimistic attitude. It's not carefully thought out at all.

The rest of the people have doubts, but are too timid to

speak out or are blocked from doing so by the Japanese corporate structure. I'm not saying there aren't resisters inside the companies. They don't make explicit objections, but they resist. For example, they might take two months to do a job that should really only take one. Or they demand stricter standards. But what happens then in most cases is that after two or three years of this kind of resistance, they are put out to pasture.

When I was a company man, I set up and led a kind of study group with some of my young colleagues. Most of the members of that group are now in unduly low positions. Some have left. But no one is in a position of power. That's the way Japanese society is.

Q6: Can't you give them support?

A: How can I possibly support them? It's quite difficult. Some important positions in Japan's nuclear power government organizations and industry, such as the chairman of the Nuclear Safety Commission of Japan, the board members of the so-and-so nuclear power committee, and the members of the special committee for the Mihama Nuclear Accident, are occupied by professors emeritus of the University of Tokyo today. These are the people in charge, the bosses of Japan's nuclear power industry. Now, how did they get to these positions of power? Anyone who makes any criticism is either fired early on or made impotent. Then there are the people like myself who choose to quit. This leaves only the yes-men on the career track. These people become the heads of academic societies. So it is not necessarily the most capable people who are given power. Some of my former classmates are now executives at their companies. If they had been in academia, they would now be professors. They would be on the investigation committee. But if you asked me if they were bright and capable and had the potential to take a position of authority back when I knew them, my answer would be, "not really." They were nothing but yes-men. I don't think they were very promising as scientists. I used to tease them for not knowing things I knew. They are now called "authorities." (Audience laugh) Authorities are literally created. The earlier you become critical, the farther away you get from authority. So it's not the way people usually imagine it to be. It's not the case that skilled, committed people are leading the nuclear power industry.

Q7: Is it possible to measure how much radiation leaked during the Mihama accident?

A: I think it would be possible to make an estimate to some extent if the proper measurement procedure was followed. But the current estimate is based on so many false premises. They got the leakage level by converting a value measured very early on. For one thing, this was a very rough conversion. For another, circumstantial evidence indicates that radiation continued to leak even after this early measurement was taken. This will probably be made public at the Diet hearings or on some other occasion, but the power company is trying to keep it hidden for now. I don't want to be too optimistic, but the truth will come out someday.

Although I cannot say anything with absolute certainty, the actual amount of leakage was probably at least ten times higher than the publicized estimate. It could be a hundred times. If they give us precise information, like what type of monitoring device was used and for how many minutes the monitoring was conducted, then we can make a better estimate. I think the publicly announced estimate of the level of radiation leakage is too low, even at this stage. I suspect that a lot of facts have been withheld.

We know that there was a guillotine rupture, which only shows that the Mihama accident was indeed a major one, much more serious than initially thought. I suspect that there are many other facts still under wraps. We must demand their disclosure. Everything the power company says is threaded with lies. They haven't given us the full story from the beginning. It was only in the face of pressure from Fukui prefectural government officials that they admitted that a relief valve did not open. Later they acknowledged that there were actually two valves instead of one and, then even later, that neither of them operated properly. There must be more they are not telling us. I think what they most want to hide is the radiation.

Q8: Who has the authority to require the submission of all the data?

A: At this stage, only the national government has. The prefectural government doesn't have such authority. The safety agreement doesn't give it investigatory powers.

Q9: What if the Diet members demand the data be submitted to the Diet?

A: The only way for this to happen is for the Diet to exercise its investigatory powers. But the government is saying that it won't release the report on the accident investigation. Why? Because it will "hinder unrestricted discussions within the Accident Investigation Committee." Do you buy this argument? What they are really saying is that if all the data were released, we could all discuss it. But then the report on the accident investigation would be meaningless. As a matter of fact, the report is prepared behind closed doors by the big brass. But if a third party got hold of the data and made a big fuss about it, then unrestricted discussions would be possible. So the Accident Investigation Committee will not release the data. Believe it or not, this kind of argument goes unchallenged. Even if it were questioned during a Diet session, the answer would never be given because time would be up. It is shameful but this is a political issue and political power dominates everything. If the data to be submitted to the committee were released, much more meaningful discussions could be held than those engaged in by the committee. In other words, if they do not keep it a secret, the committee would be rendered redundant. This is simply ridiculous. What is really disappointing to me is the fact that this is done in the name of science.

Q10: I am interested in nuclear power generation so I ask power companies lots of questions. I get the impression that they do not really want to be engaged in the nuclear power

business. So why do they not get out?

A: I do not think they are particularly aggressive, at least at the moment. There is much to be done. The nuclear waste problem in particular prevents the nuclear power business from being profitable. Accidents make matters worse. So they are not really that eager about it. So who then is promoting the nuclear power business? The government and the power plant construction sector are. The government is being lobbied by the companies that build the plants. They make good money from the nuclear power industry. Companies like Toshiba, Hitachi and Mitsubishi Heavy Industries profit from the construction of nuclear reactors. Even nuclear waste disposal can be a good business. Today, the nuclear power industry is worth around 2 to 3 trillion yen. So it is not a matter of energy or anything else. Just business. It is basically an industrial promotion measure.

Q11: Some say that if you shut down the nuclear power plants, people will lose their jobs and some may end up committing suicide. So you cannot stop the plants for that reason. How do you respond to this?

A: If there is money to invest in nuclear power, then it can be used to create other industries. In the Gulf War, nuclear reactors in Iraq were bombed, leading to radioactive contamination; this worries me. Some time after this happened, I bought a copy of TIME magazine, which carried an ad by Siemens, a German company that builds nuclear reactors. The company also produces weapons, so it is like Mitsubishi Heavy Industries in Japan. It was a full page ad about solar panels. Siemens claims that our future depends on solar power! What interested me was its argument that the solar business will absorb more labor. That is true. Nuclear power generation does not need that much labor. What was more interesting was that the ad said that the company is building the world's largest solar cell plant in Wackersdorf, West Germany. This is the city where it was supposed to build a nuclear fuel reprocessing plant. But the plan was abandoned. The ad did not mention this, but anyone who knows something about it would have noticed that instantly.

So that is the era we live in today. Using the jobs argument as a reason to continue to promote the nuclear power policy is a distraction and has nothing to do with social justice. If this argument is allowed to stand unchallenged, then we will not be able to stop any wrongdoing.

Q12: I hear that Japan is thinking about exporting its nucle-

ar power industry to China and Southeast Asia because it has hit a ceiling domestically. Is that true?

A: It has not been very successful but they are clearly making efforts to do this.

Q13: Are the plans to build nuclear power plants in other countries still on the table?

A: Building a plant requires a huge investment, so if they can get the Overseas Development Administration (ODA) funds or subsidies from the Export-Import Bank of Japan, then it is possible. Under the usual system, money is provided in the name of the government, the companies make a profit and other countries get the plants. The reason why the third world countries are so eager to have these plants is because they intend to make nuclear weapons. Japan is conducting aggressive marketing in Indonesia to get support for its plan to build plants there. It's been said that this support may soon materialize. Indonesia is rich in both oil and natural gas, so it should not need another source of energy. Then why do they want nuclear power plants so badly? Indonesia is a military dictatorship, so a large part of it is that it wants nuclear weapons. There are many other countries, like Iraq and other Middle Eastern countries, which want nuclear power plants. These countries have plenty of oil, so it's not a matter of energy. It's a matter of armaments.

Q14: I hear that Japan has committed to building nuclear power plants in China. What about that?

A: As far as China is concerned, France and the UK have the largest stakes in the market, with Japan being third. Traditionally, European countries have had stronger ties with China. But Japan is conducting its own marketing efforts. The domestic market is oversaturated, so the government may intensify its efforts in other countries.

[i] Editor's note: The **Citizens' Nuclear Information Center** is an anti-nuclear public interest organization dedicated to securing a safe, nuclear-free world. The Center was formed so as to provide reliable information and public education on all aspects of nuclear power to ultimately realize this goal. Data gathered, compiled, and analyzed by the Center is condensed into forms useful to the media, citizens' groups, policy makers, and the general public. The Center is independent of government and industry and is supported by membership fees, donations, and sales of publications.

CNIC was established in 1975 in Tokyo to collect and analyze information related to nuclear energy including safety, economic, and proliferation issues, and to conduct studies and research on such issues. During 1995-97, the center spearheaded an independent study on mixed uranium-plutonium oxide (MOX) fuel with a number of prominent researchers from Europe, the U.S.A., and Japan. The Research Director of the International MOX Assessment (IMA) project, Dr. Jinzaburo Takagi, together with the Research Sub-Director, Mycle Schneider, received the Right Livelihood Award for this study and activities relating to plutonium issues. The results of this project have been published in Japanese, English, French, and Russian.

Citizens' Nuclear Information Center
Akebonobashi Co-op 2F-B
8-5 Sumiyoshi-cho, Shinjuku-ku,
Tokyo 162-0065, JAPAN
Twitter: @CNIC.Japan
http://www.ustream.tv/channel/cinic-news

Tel: 03-3357-3800

FAX: 03-3357-3801

[ii] Editor's note. The Great East Japan Earthquake occurred off the Pacific coast of the Tohoku region affecting the East and Northeastern areas of Japan with a huge tsunami on 11 March, 2011. This tsunami caused a number of nuclear accidents, primarily level 7 meltdowns at three reactors in the Fukushima Daiichi Nuclear Power Plant complex, and the associated evacuation zones affecting hundreds of thousands of residents.

[iii] Editor's note: See note 11.

[iv] Editor's note: The **Gulf War** (2 August 1990 - 28 February 1991) was a war waged by a United Nations-authorized coalition force consisting of 34 nations led by the United States, against Iraq, in response to Iraq's invasion and annexation of Kuwait. The present lecture was given on 22 February 1991, while the war still raged on.

[v] Editor's note. See Appendix for the list of Jinzaburo Takagi's books.

[vi] Editor's note. The most important issue for the safety of the nuclear power plant is the prevention of nuclear leakage. The counter measure taken for this is the quintuple-barrier structure. It consists of five barriers for nuclear leakage prevention: 1) the nuclear fuel pellet, 2) the nuclear fuel cladding tube,

3) the reactor pressure vessel, 4) the reactor container, and 5) the reactor building.

[vii] Editor's note. See Jinzaburo Takagi,『科学は変わる』(Science is Changing) p. 13.

[viii] Editor's note. Here, Jinzaburo Takagi refers to the historical transformation of knowledge that occurred in the 16th and 17th centuries in Europe, and lays out the foundation of what we generally understand as modern scientific knowledge. Thanks to this historical transformation, today it is generally believed that modern science is unique to Europe, and that it is a product of European modernity.

The branch of knowledge that played the central role in this historical transformation was pure mathematics. Prior to this era and in many other parts of the world - ancient Greece, China, India, and so forth - people had known of the systematic nature of mathematical knowledge, and it had long been compiled and institutionalized. But the role of mathematics underwent a change in relation to the other disciplines of knowledge in such a radical way in early modernity that the forms of scientific knowledge could never return to those previous forms. This change was propelled by a great number of historical accomplishments including Galileo's mathematization of nature, Descartes' introduction of the orthogonal coordinate system, Newton's and Leibniz' methods of infinitesimal calculus, and so forth. Scientific knowledge was reorganized and formalized in terms of the elaborate notational system of mathematical equations, and mathematics helped scientists isolate the realm of complete and absolute abstraction in their thinking. Precisely because mathematics teaches nothing about their experience can it allow them to separate rational and conceptual coherence from empirical discoveries in their reasoning. The domain of experience could thus be isolated from the domain of reasoning, which was explicitly articulated and formalized by mathematics. Thanks to the mathematization of knowledge, modern science has gained an unprecedented rigor of methodological consciousness.

[ix] Editor's note: On nuclear stability. The nucleus of a radioisotope is unstable. In an attempt to reach a more stable arrangement of its protons and neutrons, the nucleus will spontaneously decompose to form a different nucleus. If the number of neutrons changes in the process, a different isotope is formed. If the number of protons changes in the process, then an atom of a different element is formed. This decomposition of the nucleus is referred to as radioactive decay. During this decay an unstable nucleus spontaneously decomposes to form a different nucleus, giving off radiation in the form of atomic particles or high-energy rays. This decay occurs at a constant, predictable rate referred to as half-life. A stable nucleus will not undergo this kind of decay and is thus non-radioactive.

[x] Editor's note. I suggest that some comment about Takagi's tropic strategy should be included. The question of nuclear stability urged him to use the trope of "gods' fire," which he asserted repeatedly, cannot be found on earth.

Therefore, he says that life cannot live near gods' fire. Somewhat this reference to gods' fire is absent in the middle part of the transcription, where it could have been most effective. I suspect that the initial transcription was not done properly, so that the Japanese edition did not pay attention to the tropic organization of the entire lecture.

[xi] Editor: The accident which Jinzaburo Takagi refers to occurred on 2nd September, 1991: it happened at the Mihama No. 2 nuclear power station in Fukui prefecture, Japan. This plant is owned and run by Kansai Electric Power Corporation, the major power utility in Western Japan. One tube in Unit 2 steam generator was completely broken. This triggered an emergency shutdown of a nuclear reactor with full activation of the Emergency Core Cooling System. Eventually, a small amount of radiation was released to the outside.

An accident of a similar nature occurred on 9 August 2004. Four workers were scalded to death by superheated steam, and seven others were injured. This accident was caused by a burst steam pipe in the non-radioactive part of the reactor. In 27 years of operation the steam pipe had never been checked once for corrosion, let alone replaced. By the time it burst, its walls had worn down from an initial 10 mm of carbon steel to a mere 1.4 mm. Nine months before the accident a subcontractor company had alerted the operators to the need for inspections, but the warning was ignored.

The Mihama No. 3 reactor, which started commercial operations in 1976, is of the pressurized water (PWR) design used in the US, Germany, and France. High-pressure water is used to carry heat from the reactor core to a heat exchanger. There its heat turns water of a secondary cooling cycle into steam, which then drives power turbines. There are 15 nuclear power stations in Fukui prefecture, almost one third of the 47 in all of Japan.

This accident raises serious renewed questions about the safety culture of Japanese Nuclear Power Plant operators and the legal framework that surrounds them. Existing regulations did not explicitly require the company to check the secondary cooling cycle steam pipes of nuclear power stations. In 1995 a serious accident forced a shutdown in the Monju Fast Breeder Reactor, also in Fukui prefecture. That accident was also caused by a broken cooling pipe. The operators then tried to conceal evidence about the magnitude of the accident.

[xii] Editor's note. "*Shi no hai* (death ashes)" was a phrase invented in Japan to describe **nuclear fallout** in the event of the *Daigo Fukuryū Maru* (*the Lucky Dragon #5*), a Japanese fishing boat that encountered the fallout from the Castle Bravo nuclear test conducted by the United States, at Bikini Atoll near the Marshall Islands, on 1 March 1954. The boat with its 23 fishermen were contaminated. The crew members were diagnosed with acute radiation syndrome. On September 23, the chief radio operator, Aikichi Kuboyama, 40, died and became the first Japanese victim of a hydrogen bomb.

It was reported that the sky in the west lit up like a sunrise. Eight minutes later the sound of the explosion arrived, with fallout coming several hours later. The fallout had absorbed highly radioactive products. This fell on the boat for three hours. After the radiation sickness symptoms appeared, the fishermen called it shi no hai (**death ash**). The United States government refused to disclose its composition due to "national security." The head of the United States Atomic Energy Commission issued a series of denials; he also suggested that the lesions on the fishermen's bodies were not caused by radiation; that they were inside the danger zone (while they were 40 miles away); and told President Eisenhower's press secretary that the Lucky Dragon Number Five may have been a "Red spy outfit", commanded by a Soviet agent. He also denied the extent of the claimed contamination of the fish caught by Daigo Fukuryu Maru and other ships in the vicinity.

This incident marked the beginning of the largest anti-nuclear citizens' movement in Japan since the end of the Asia-Pacific War.

[xiii] Editor's note: **Half-life** (**t½**) is the time required for a quantity to fall to half its value as measured at the beginning of the time period. In physics, it is typically used to describe a property of radioactive decay, but may be used to describe any quantity that follows an exponential decay. The term itself hinges upon Ernest Rutherford's discovery of the principle in 1907.

[xiv] Editor's note: The **Kashiwazaki-Kariwa Nuclear Power Plant** is a large nuclear power plant owned and operated by The Tokyo Electric Power Company, in the towns of Kashiwazaki and Kariwa in Niigata Prefecture, Japan. It is located near the Sea of Japan, from where it gets cooling water.
It was the largest nuclear generating station in the world by net electrical power rating when it started operating in 1997. But the plant was severely af-

fected by earthquakes. In July 2007, it was approximately 15 miles from the epicenter of the second strongest earthquake, the Mw 6.6, to ever occur at a nuclear plant. This shook the plant beyond designed safety levels and initiated an extended shutdown for inspection; this indicated that greater earthquake-proofing was needed before operation could be resumed. The plant was completely shut down for 21 months following this earthquake.

Against much protest, the Tokyo Electric Power Company restarted Unit 7 after seismic upgrades on May 9, 2009, followed later by units 1, 5, and 6. (Units 2, 3, 4 were not restarted). However all units have since been stopped for regular inspection since the Great Eastern Japan Earthquake of 11 March, 2011.

[xv] (Editor's note) The Rokkasho village in Aomori prefecture in northern Japan is the place where it is planned to keep the radioactive substances from all the nuclear power plants in Japan, Jinzaburo Takagi wrote the book Critique of the Nuclear Recycle Plant [Nanatsumori Shobo, 1991] on this topic. And *Testimony: the Future of the Nuclear Recycle Plant* [Nanatsumori Shobo, 2000] (『核燃料サイクル施設批判』1991年，七つ森書房，『証言核燃料サイクル施設の未来は』2000年，七つ森書房]. So, the Rokkasho plant is meant not only for processing of radioactive substances from the Mihara nuclear plant, but also from all the other nuclear plants in Japan. As a matter of fact, some of radioactive substances were transported to France for processing under a nuclear treatment treaty between Japan and France.

[xvi] Editor’s note: see note 11 above.

[xvii] Editor’s note:『巨大事故の時代』（弘文堂，1989），The Era of Catastrophic Disaster, Kôbun-dô, 1989）

[xviii] Editor’s note: see note 4 above

[xix] Editor’s note: Kenji **Miyazawa**（**1896 - 1933**）was a Japanese poet and author of children's literature in the early Shōwa period in Japan. He was also known as a devout Buddhist, vegetarian and social activist. He wrote experimental works in poetry and children’s literature in which a new cosmology and socialist ideals werepresented. Undoubtedly his poetry marks a zenith in Japanese modernity poetry, and his literature marks the highest achievement in Japanese children’s literature. His major accomplishments include *Spring and Asura*（春と修羅）, *Nights on the Galactic Railroad*（銀河鉄道の夜）, *Matasaburo of the Wind*（風の又三郎）, *The Restaurant of Many Orders*（注文の多い料理店）and *the Okhotsk Elegy*（オホーツク挽歌）. In works such as “The Life of Guskov Voudry（グスコーブドリの伝記）”, a vision of social alliance between intellectuals and peasantry is projected. In his hometown of Hanamaki, he continued to work as a social activist until his premature death.

[xx] Editor’s note: See note 1 above.

[xxi] Editor's note: 高木仁三郎・渡辺美紀子,『食卓にあがった放射能』東京，七つ森書館，2011（*Radioactivity on the Dining Table*, Tokyo, Nanatsumori Shokan, 2011）

◇著者・訳者略歴

高木仁三郎（たかぎ じんざぶろう）

1938（昭和13）年群馬県生まれ。61年東京大学理学部卒業。その後、日本原子力事業、東京大学原子核研究所などを経て、75年原子力資料情報室の設立に参加し、86年より同代表（〜98）。この間、プルトニウム利用問題の批判的研究と活動で国際的に評価を得る。97年ライト・ライブリフッド賞（Right Livelihood Award）ほか多数の賞を受賞。原子力時代の末期症状による大事故の危険性と放射性廃棄物がたれ流しになっていることに危惧の念を“最後のメッセージ”に残し、2000（平成12）年に没。

著書は、『プルトニウムの恐怖』（岩波新書、1981）、『市民科学者として生きる』（同、99）、『原発事故はなぜくりかえすのか』（同、2000）、『高木仁三郎著作集』全12巻（七つ森書館、01〜04）、『いま自然をどうみるか』（新装版、白水社、11）ほか多数。

酒井直樹（さかい なおき）

1946（昭和21）年神奈川県生まれ。71年東京大学文学部卒業。83年シカゴ大学人文学部で博士号取得。現在、コーネル大学ゴールドウィン・スミス記念基金教授。専門は比較文学・比較思想史・翻訳理論、人種主義・国民主義研究など。著書は、『日本思想という問題』（岩波書店、1997）、『過去の声』（日本語版、以文社、2003）、『日本・映像・米国』（青土社、2006）、『希望と憲法』（以文社、2008）、『死産される日本語・日本人』（新曜社、1995、のち講談社学術文庫、2015）、『パックス・アメリカーナの終焉とひきこもりの国民主義』（岩波書店、近刊）など多数。

The Law of Science versus the Law of Life
Nuclear Accidents and the Limits of Human Control
（日本語版『科学の原理と人間の原理　人間が天の火を盗んだ
—その火の近くに生命はない』）

2015年11月16日　初版第1刷発行

著　者　高木 仁三郎
訳　者　酒井直樹
発行者　光本　稔
発　行　株式会社　方丈堂出版
〒601-1422　京都市伏見区日野不動講町38-25
電話　075-572-7508　FAX　075-571-4373
発　売　株式会社　オクターブ
〒606-8156　京都市左京区一乗寺松原町31-2
電話　075-708-7168　FAX　075-571-4373
装　幀　小林　元

印刷・製本　亜細亜印刷㈱

ISBN　978-4-89231-136-9